COLLECT-A-PET READER

My cute

kitten

Written by Helen Anderton

make
believe
ideas

Reading together

This book is an ideal first reader for your child, combining simple words and sentences with beautiful photography of kittens. Here are some of the many ways you can help your child with their early steps in reading.

Encourage your child to:

- Look at and explore the detail in the pictures.
- Sound out the letters in each word.
- Read and repeat each short sentence.

Look at the pictures

Make the most of each page by talking about the pictures and spotting key words. Here are some questions you can use to discuss each page as you go along:

- Do you like this kitten?
- If so, what do you like about it?
- What would it feel like to touch?
- How would you look after it?

Look at rhymes

Many of the sentences in this book are simple rhymes. Encourage your child to recognise rhyming words. Try asking the following questions:

- What does this word say?
- Can you find a word that rhymes with it?
- Look at the ending of two words that rhyme. Are they spelt the same? For example, "day" and "play", "chase" and "pace".

Test understanding

It is one thing to understand the meaning of individual words, but you need to check that your child understands the facts in the text.

- Play "spot the mistake". Read the text as your child looks at the words with you, but make an obvious mistake to see if he or she has understood. Ask your child to correct you and provide the right word.
- After reading the facts, shut the book and make up questions to ask your child.
- Ask your child whether a fact is true or false.
- Provide your child with three answers to a question and ask him or her to pick the correct one.

Kitten quiz

At the end of the book, there is a simple quiz. Ask the questions and see if your child can remember the right answers from the text. If not, encourage him or her to look up the answers.

Key words

These pages provide practice with very common words used in the context of the book. Read the sentences with your child and encourage him or her to make up more sentences using the key words listed around the border.

Picture dictionary

A picture dictionary page illustrates the things you need when looking after a kitten.

Watch me grow

I'm a tiny ball of fur.
Soon I'll learn to paw and purr.
Until then, I'll dream all day
of being big enough to play!

tail

DID YOU KNOW?

When kittens are born, they can't see or hear.

eye

Mummy and me

For several days, I cannot see,
but Mummy stays right next to me.
She keeps me warm
and gives me milk,
and licks me so
my fur's like silk.

Mummy

DID YOU KNOW?

Newborn kittens can't control their body temperature, so they must keep close to their mother to stay warm.

silky fur

Learning to walk

Look at how the time has flown –
I'm learning how to walk alone!
I'm feeling very brave and bold
now I'm nearly four weeks old.

leg

DID YOU KNOW?

Kittens begin learning to walk when they are three weeks old.

paw

9

Time to eat

As I grow, I learn to eat
a different kind of kitten treat:
yummy wet food from a tin,
with meat or fish or jelly in!

DID YOU KNOW?
Kittens grow 15 times faster than babies, and eat up to six meals a day!

wet food

bowl

Scratch!

Do you see what's on my paws?
I've got a set of long, sharp claws!
I like to scratch the walls and floors,
and your curtains, clothes and doors!

claw

paw

DID YOU KNOW?
Kittens scratch things to sharpen their claws and to mark their territory.

scratching post

Play with me

Would you like to play with me?
I've got so much energy!
Pull some wool across the ground –
I'll run and chase it round and round!

DID YOU KNOW?

Kittens will chase a piece of wool because it looks like a mouse's tail.

wool

15

Purr!

When I'm sad, I like to mew
and wave my fluffy tail at you.
But if you gently
stroke my fur,
I will smile and
give a purr!

DID YOU KNOW?

Kittens can make around 60 different sounds. Each sound means something different.

fur

17

Zzzzzzzz!

When I'm tired, I like to nap
in my bed or on your lap.
Curled up in a fuzzy ball,
I'm furry, warm and very small!

DID YOU KNOW?

Kittens need 16–20 hours of sleep a day. This helps them to save energy for playtime!

Hiding places

My favourite kind of hiding place
is a warm and cosy space:
in a box, under the bed,
or tucked up in your clothes, instead!

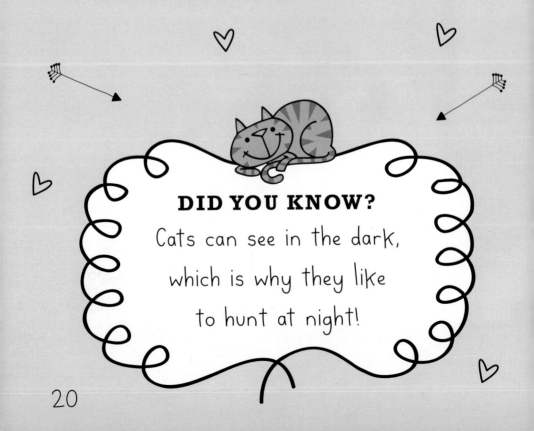

DID YOU KNOW?
Cats can see in the dark,
which is why they like
to hunt at night!

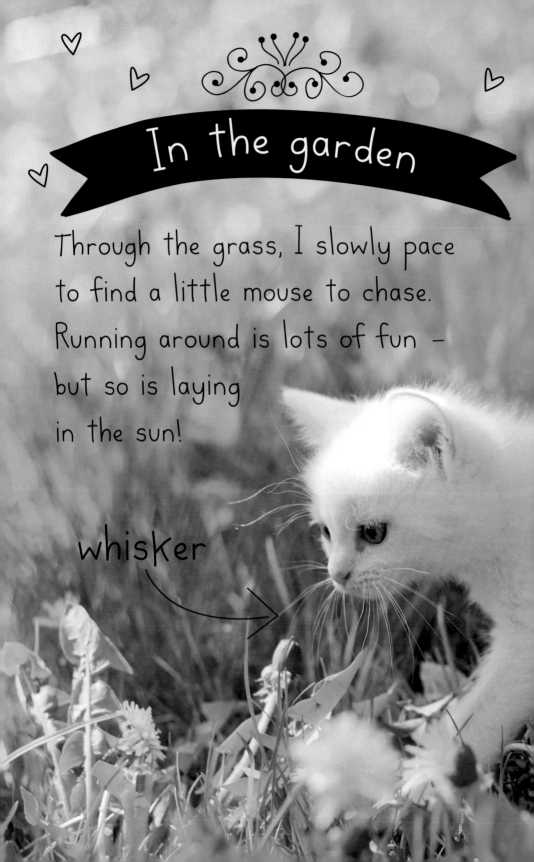

In the garden

Through the grass, I slowly pace
to find a little mouse to chase.
Running around is lots of fun –
but so is laying
in the sun!

whisker

DID YOU KNOW?

Kittens use their whiskers for balance and to feel what is around them.

Let's visit the vet

It's time to go and see the vet.
She makes sure I'm a healthy pet.
I'll come back from time to time,
to keep me well and feeling fine.

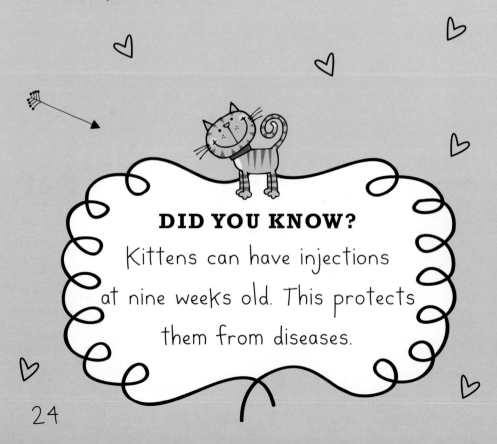

DID YOU KNOW?
Kittens can have injections
at nine weeks old. This protects
them from diseases.

vet

25

All grown up

Now I'm grown up – look at me!
I'm big and strong, as you can see.
But I still like to run and play,
and spend time with you every day!

DID YOU KNOW?

Most kittens don't reach their full size until they are 18 months old.

Kitten quiz

How much do you know about me?

1. How many sounds can kittens make?

They can make around 60 sounds.

2. Why do kittens scratch things?

They scratch things to sharpen their claws and to mark their territory.

3. How many meals can kittens eat a day?

They can eat up to six meals a day.

4. Can kittens see
or hear when they
are born?

No.

5. At what age
do kittens reach
their full size?

They reach their full size at 18 months old.

6. What do kittens
use their whiskers for?

They use them for balance and
to feel what is around them.

7. Can cats
see in the dark?

Yes.

I ♡ up ♡ look ♡ we ♡ like ♡ and ♡

yes ♡ it ♡ see ♡ she ♡ me ♡ of ♡ in ♡ come ♡

Key words

Here are some key words used in context. Help your child to use other words from the border in simple sentences.

I have a rough tongue.

I **play** with my toy.

cat ♡ am ♡ all ♡ mummy ♡ my ♡

on ♡ at ♡ for ♡ a ♡ he ♡ is ♡ go

Look at my soft paws.

I sleep in the day.

My whiskers are long.

I like to eat meat.

you ♡ are ♡ this ♡ going ♡ they ♡ away ♡ play

big ♡ dog ♡ the ♡ day ♡ get can ♡

Picture dictionary

bed

bowl

brush

carrier

cat food

collar

mouse

scratching post

toy